AI MEETS QUANTUM

Synergies and Future Applications

Najeed Khan

CONTENTS

Title Page

Preface

Prologue

Chapter 1: Introduction: The Convergence of AI and Quantum Computing   1

Chapter 2: Quantum Computing for AI Optimization   7

Chapter 3: Quantum Machine Learning   13

Chapter 4: AI for Quantum System Design   21

Chapter 5: Applications in Natural Sciences   29

Chapter 6: Revolutionizing Cryptography   37

Chapter 7: Quantum-AI Fusion in Finance   45

Chapter 8: Healthcare and Drug Discovery   53

Chapter 9: AI-Enhanced Quantum Networking   61

Chapter 10: Ethics and Challenges in Quantum-AI Systems   69

Bibliography   77

About The Author   81

Books By This Author   85

PREFACE

The year 2025 has been designated as the International Year of Quantum (IYQ 2025), a global celebration of the profound advancements and transformative potential of quantum science and technology. As humanity stands at the frontier of a quantum revolution, it is a fitting moment to explore how this paradigm shift intersects with another groundbreaking force: artificial intelligence (AI). *AI Meets Quantum: Synergies and Future Applications* is born from the belief that the convergence of these two domains is not only inevitable but also essential for addressing some of the most pressing challenges of our time.

The journey of this book began with a simple yet profound question: What happens when two of the most transformative technologies of the 21st century come together? Over the past decade, quantum computing has transitioned from a theoretical curiosity to a practical tool with immense computational power, while AI has become an indispensable driver of innovation, permeating every aspect of our lives. The intersection of these fields is a fertile ground for discovery, promising advances that could redefine industries, accelerate scientific breakthroughs, and reshape the global economy.

This book is designed to serve as a comprehensive guide to the emerging synergies between AI and quantum computing. It explores how quantum computing can optimize AI algorithms, how AI can enhance quantum system design, and how their fusion is unlocking unprecedented applications across domains such as natural sciences, finance, healthcare, and cryptography. Through ten carefully curated chapters, we delve into the technical underpinnings, real-world applications, and societal implications of this convergence. Each chapter draws on leading research, key case studies, and insights from pioneers in both fields, providing a balanced perspective that bridges the technical and the visionary.

As we celebrate IYQ 2025, it is also imperative to reflect on the responsibilities that come with such transformative power. The final chapter of this book addresses the ethical challenges, societal impacts, and technical hurdles associated with quantum-AI systems. By acknowledging these complexities, we hope to contribute to a thoughtful and inclusive dialogue about the future we are collectively shaping.

This book is intended for a diverse audience: technologists and researchers eager to explore the technical intricacies of quantum-AI systems; industry leaders and policymakers seeking to understand their transformative potential; and curious readers who wish to grasp the profound implications of this convergence. Whether you are an expert or an enthusiast, we invite you to join us on this exploration of one of the most exciting frontiers in science and technology.

We are living in a time of unprecedented possibility, where the boundaries between imagination and reality are constantly shifting. As you turn these pages, we hope to inspire not only a deeper understanding of AI and quantum computing but also a sense of wonder and responsibility for the future they will help create.

Welcome to *AI Meets Quantum: Synergies and Future Applications*. May this book spark your curiosity, challenge your thinking, and ignite your imagination.

Author's Note: This book is a product of collaboration, curiosity, and a deep respect for the transformative potential of science and technology. I am immensely grateful to the researchers, engineers, and visionaries whose work has paved the way for this exploration. To the readers: thank you for embarking on this journey with me. Let us continue to push the boundaries of what is possible.

PROLOGUE

AI Meets Quantum: Synergies and Future Applications delves into the groundbreaking intersection of artificial intelligence (AI) and quantum computing, exploring how these two transformative technologies are reshaping industries and advancing human potential. The book offers a comprehensive overview of their convergence, highlighting both opportunities and challenges while showcasing their profound implications across a range of domains.

Key Themes and Chapters

1. **The Convergence of AI and Quantum Computing**
   The book begins by framing the synergy between AI and quantum computing, presenting a visionary outlook on how these technologies complement each other in reshaping the technological landscape.

2. **Quantum Computing for AI Optimization**
   It examines how quantum computing enhances AI systems, optimizing model training, algorithms, and computational processes through methods such as the Quantum Approximate Optimization Algorithm (QAOA).

3. **Quantum Machine Learning**
   The book explores hybrid quantum-classical approaches and native quantum machine learning techniques, offering insights into their applications and potential to transform

traditional AI paradigms.

4. **AI for Quantum System Design**
   AI's role in designing quantum systems is highlighted, focusing on how it aids quantum hardware development, automates quantum experiments, and improves error correction mechanisms.

5. **Applications in Natural Sciences**
   The synergy of AI and quantum computing in chemistry, biology, and physics is discussed, emphasizing breakthroughs in material discovery, molecular simulation, and fundamental scientific research.

6. **Revolutionizing Cryptography**
   The book investigates the dual-edged impact of quantum-AI systems on cryptography, detailing how they advance secure data encryption while also posing risks to classical cryptographic methods.

7. **Quantum-AI Fusion in Finance**
   It uncovers applications in risk analysis, portfolio optimization, and fraud detection, showcasing how quantum-AI systems are reshaping the financial industry.

8. **Healthcare and Drug Discovery**
   The potential of quantum-AI technologies in accelerating drug discovery, genomic analysis, and personalized medicine is examined, with real-world examples of their transformative impact.

9. **AI-Enhanced Quantum Networking**
   AI's contributions to quantum communication and the development of secure quantum networks are discussed, emphasizing the journey toward scalable and reliable quantum internet infrastructure.

10. **Ethics and Challenges in Quantum-AI Systems**

The final chapter tackles ethical concerns, societal impacts, and technical challenges at the quantum-AI interface, stressing the need for collaboration, regulation, and responsible innovation.

Contribution to the Field

AI Meets Quantum offers an interdisciplinary perspective on the convergence of AI and quantum computing, blending technical insights with broader societal implications. Drawing on key references and groundbreaking research, the book serves as a guide for technologists, researchers, and policymakers navigating this rapidly evolving landscape.

Vision for the Future

The book concludes with a call to action for fostering innovation while addressing ethical and technical challenges. By leveraging the unique strengths of AI and quantum computing, humanity has the opportunity to solve complex problems and usher in a new era of scientific and societal progress.

This work is a must-read for those at the forefront of technology and for anyone interested in understanding the profound possibilities and responsibilities associated with quantum-AI systems.

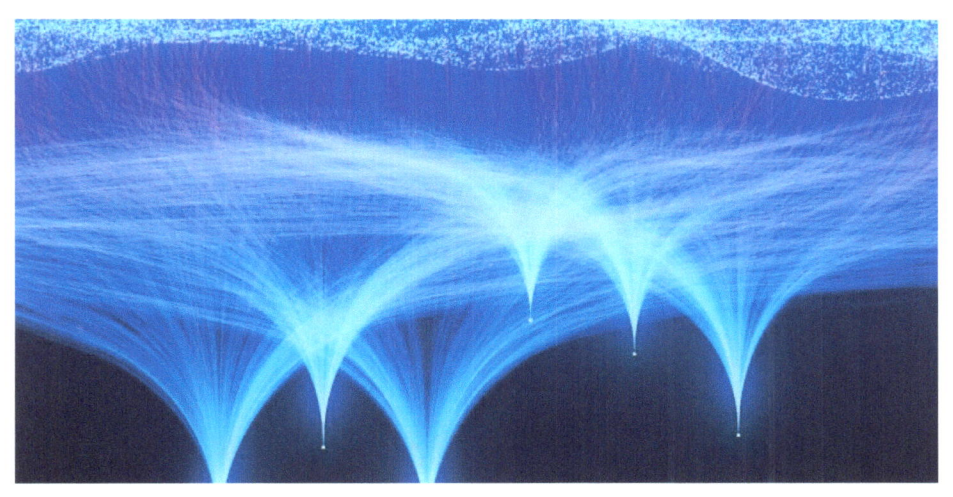

CHAPTER 1: INTRODUCTION: THE CONVERGENCE OF AI AND QUANTUM COMPUTING

In the annals of technological evolution, few intersections hold as much transformative potential as the convergence of artificial intelligence (AI) and quantum computing. Both disciplines are not only redefining their respective fields but are also unlocking unprecedented possibilities when combined. This chapter sets the stage for exploring how these groundbreaking technologies are reshaping the technology landscape and driving innovation across industries.

The Unparalleled Promise of AI and Quantum Computing

AI, as explored in Tegmark's *Life 3.0* (2017), has emerged as a cornerstone of modern innovation. From natural language processing to autonomous systems, AI has demonstrated the ability to solve complex problems, learn from vast datasets, and make predictions with uncanny accuracy. The exponential growth in computational power—coupled with sophisticated algorithms—has propelled AI to the forefront of global technological progress.

Quantum computing, on the other hand, challenges the very foundation of classical computation. As described by Nielsen and Chuang in *Quantum Computation and Quantum Information* (2010), quantum computers leverage the principles of superposition, entanglement, and interference to perform calculations that would be infeasible for classical systems. Unlike classical bits, quantum bits (qubits) can exist in multiple states simultaneously, enabling parallel computations that exponentially accelerate problem-solving for specific tasks.

A Synergistic Transformation

When AI meets quantum computing, the result is not merely an additive combination but a synergistic transformation. Quantum computing's ability to process massive amounts of data in parallel and solve optimization problems aligns seamlessly with AI's data-driven methodologies. Together, they address critical bottlenecks in computation, memory, and processing speed.

For example, training large-scale AI models is a resource-intensive endeavor, often limited by classical hardware constraints. Quantum computing offers the potential to optimize neural network architectures, accelerate training processes, and improve energy efficiency. Similarly, AI can enhance quantum computing by automating error correction, improving quantum gate fidelity, and designing novel quantum algorithms. This interplay not only advances each field but also catalyzes applications that were previously unimaginable.

Disrupting Industries and Redefining Possibilities

The convergence of AI and quantum computing is not a distant vision; it is already reshaping industries. In finance, quantum-AI systems are revolutionizing risk analysis, portfolio optimization, and fraud detection. In healthcare, they are expediting drug discovery and enabling precision medicine. The natural sciences, particularly chemistry and physics, are witnessing breakthroughs in molecular modeling and simulations.

Moreover, this convergence is poised to tackle humanity's grand challenges. From combating climate change through optimized energy solutions to advancing cryptographic security in the post-quantum era, the fusion of AI and quantum computing offers a toolkit for addressing the most pressing issues of our time.

Challenges and Ethical Considerations

Despite its promise, the convergence of AI and quantum computing is not without challenges. Technical hurdles, such as maintaining qubit coherence and scaling quantum systems, persist. Additionally, ethical considerations, as emphasized by Tegmark (2017), must be at the forefront. Issues of algorithmic bias, data privacy, and the societal impact of such powerful technologies require proactive governance and interdisciplinary collaboration.

Conclusion

The convergence of AI and quantum computing represents a pivotal moment in the trajectory of technological innovation. By harnessing the unique strengths of each field, we stand on the brink of a new era—one defined by unparalleled computational capabilities and transformative applications. As this book will explore, the synergies between AI and quantum computing are not merely theoretical but are already shaping the future of industries, science, and society.

Let us now delve deeper into the mechanics and applications of this convergence, beginning with an exploration of how quantum computing can optimize AI systems.

AI MEETS QUANTUM

NAJEED KHAN

CHAPTER 2: QUANTUM COMPUTING FOR AI OPTIMIZATION

As artificial intelligence continues to revolutionize industries, the demand for efficient and scalable computational methods intensifies. Quantum computing emerges as a game-changer in this landscape, offering unprecedented capabilities to optimize AI models, training processes, and algorithms. This chapter explores the symbiotic relationship between quantum computing and AI, focusing on leveraging quantum technologies to overcome computational bottlenecks and enhance AI systems.

The Role of Quantum Computing in Optimization

Optimization lies at the heart of AI. From training neural networks to fine-tuning hyperparameters, AI systems rely on solving high-dimensional optimization problems. Classical approaches often struggle with the computational complexity and resource demands of these tasks, particularly as models grow larger and more intricate.

Quantum computing, with its unique principles of superposition and entanglement, offers a paradigm shift in addressing these challenges. Farhi et al. (2014) introduced the Quantum Approximate Optimization Algorithm (QAOA), a framework designed to tackle combinatorial optimization problems. By encoding the problem into a quantum system, QAOA iteratively refines solutions, leveraging quantum parallelism to explore multiple possibilities simultaneously. This approach has significant implications for AI, enabling faster and more efficient optimization of models and algorithms.

Accelerating Neural Network Training

Training AI models, particularly deep neural networks, is a resource-intensive process requiring extensive computational power. Quantum computing provides opportunities to accelerate this process through quantum-enhanced optimization techniques. For instance, hybrid quantum-classical algorithms combine the strengths of both paradigms: classical systems handle large-scale data processing, while quantum systems perform specific optimization tasks.

One promising approach involves using quantum gradient descent methods to optimize neural network weights. These methods exploit the probabilistic nature of quantum computing to escape local minima more effectively than classical algorithms. Schuld and Killoran (2019) demonstrated the potential of quantum machine learning in feature Hilbert spaces, where quantum systems map classical data into high-dimensional representations, enabling more efficient training and better generalization of models.

Enhancing Model Architecture Search

The design of AI architectures, such as convolutional or transformer networks, often involves a trial-and-error process known as neural architecture search (NAS). This process can be computationally expensive and time-consuming. Quantum computing introduces novel strategies for exploring the search space more efficiently.

Quantum-inspired algorithms, including QAOA, can identify optimal configurations of layers, activation functions, and connectivity patterns. By encoding the NAS problem into a quantum system, these algorithms rapidly evaluate multiple candidate architectures, guiding researchers toward more effective designs with reduced computational overhead.

Improving Data Representations

Feature extraction and representation learning are critical components of AI systems. Quantum computing enhances these processes by enabling the creation of quantum feature spaces, as discussed by Schuld and Killoran (2019). These spaces allow classical data to be embedded into a quantum state, where complex relationships and patterns can be uncovered more efficiently.

Quantum feature spaces are particularly advantageous for applications involving high-dimensional data, such as image recognition, natural language processing, and genomics. By leveraging quantum properties, AI models can achieve higher accuracy and robustness in extracting meaningful insights from data.

Overcoming Current Challenges

While the potential of quantum computing for AI optimization is immense, significant challenges remain. Quantum hardware is still in its nascent stages, with issues such as limited qubit coherence, noise, and error rates posing obstacles to practical implementation. Additionally, the development of quantum algorithms tailored for AI applications requires interdisciplinary expertise and ongoing research.

Despite these hurdles, hybrid quantum-classical systems offer a viable path forward. By integrating quantum processors as specialized accelerators within classical AI workflows, researchers can harness the strengths of both paradigms while mitigating current limitations.

Conclusion

The intersection of quantum computing and AI represents a frontier of innovation, with quantum technologies poised to transform how we optimize AI models, training processes, and algorithms. From accelerating neural network training to enhancing feature representation and architecture search, quantum computing's capabilities offer a glimpse into the future of AI. As we navigate the challenges and opportunities of this convergence, the insights from pioneers such as Farhi et al. (2014) and Schuld and Killoran (2019) will guide the way toward a new era of computational excellence.

In the next chapter, we will delve deeper into quantum machine learning, exploring how hybrid quantum-classical approaches and quantum-native algorithms are shaping the evolution of AI.

NAJEED KHAN

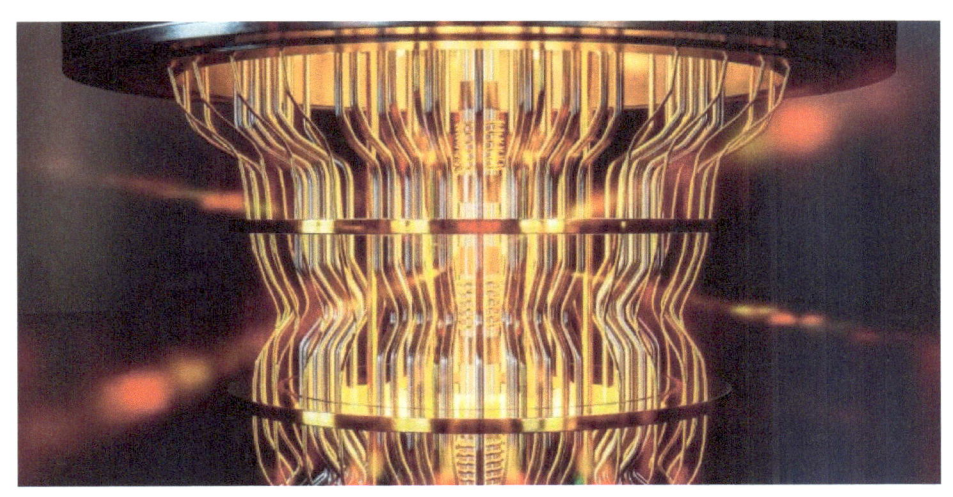

CHAPTER 3: QUANTUM MACHINE LEARNING

The convergence of quantum computing and artificial intelligence has catalyzed the emergence of quantum machine learning (QML), a field that promises to redefine the capabilities of AI systems. By blending quantum mechanics with machine learning, QML offers novel approaches to data analysis, optimization, and pattern recognition. This chapter explores the two primary dimensions of QML: hybrid quantum-classical approaches and quantum-native machine learning algorithms, drawing insights from the foundational work of Biamonte et al. (2017) and Schuld and Petruccione (2018).

Hybrid Quantum-Classical Approaches

Hybrid quantum-classical models represent a pragmatic and powerful avenue for integrating quantum computing into AI workflows. These approaches combine the computational strengths of quantum systems with the flexibility and scalability of classical systems. In this paradigm, quantum processors serve as accelerators for specific tasks, such as feature mapping, kernel evaluation, or solving optimization problems, while classical systems handle data preprocessing and model evaluation.

Quantum Kernel Methods

Kernel methods, widely used in classical machine learning, are enhanced by quantum computing through the concept of quantum kernels. As described by Schuld and Petruccione (2018), quantum kernels map classical data into high-dimensional quantum Hilbert spaces, enabling the detection of intricate patterns that are otherwise inaccessible to classical algorithms. By leveraging quantum superposition and entanglement, these methods achieve superior performance in tasks such as classification and clustering.

Quantum-enhanced support vector machines (QSVMs) are a notable example of this hybrid approach. In QSVMs, quantum circuits calculate kernel matrices, significantly reducing computational complexity and improving the accuracy of predictions.

Variational Quantum Algorithms

Variational quantum algorithms (VQAs) are another cornerstone of hybrid quantum-classical machine learning. VQAs optimize parameterized quantum circuits (PQCs) using classical optimization techniques. Examples include the Variational Quantum Eigensolver (VQE) and the Quantum Approximate Optimization Algorithm (QAOA), both of which have applications in feature selection, neural network training, and unsupervised learning.

By iteratively tuning the parameters of PQCs, VQAs enable the efficient exploration of high-dimensional solution spaces, offering a path to solve complex machine learning problems with quantum hardware that is accessible today.

Quantum-Native Machine Learning Algorithms

Quantum-native machine learning algorithms are designed to operate entirely within the quantum paradigm, leveraging the unique properties of quantum systems to perform tasks that are challenging or impossible for classical methods. These algorithms exploit quantum principles such as superposition, entanglement, and quantum interference to create fundamentally new ways of learning from data.

Quantum Neural Networks

Quantum neural networks (QNNs) represent a quantum-native approach to replicating the functionality of classical neural networks. As discussed by Biamonte et al. (2017), QNNs encode information into quantum states and perform operations using quantum gates. These networks have the potential to achieve exponential speedups in tasks such as image recognition, natural language processing, and reinforcement learning.

One promising model is the quantum perceptron, which mimics the functionality of a classical perceptron but operates entirely within a quantum system. Quantum perceptrons leverage quantum parallelism to process multiple inputs simultaneously, enhancing their capacity for complex data representation.

Quantum Boltzmann Machines

Quantum Boltzmann machines (QBMs) extend the classical Boltzmann machine by utilizing quantum states to represent probability distributions. By exploiting quantum tunneling, QBMs overcome the limitations of classical systems in escaping local minima, resulting in more efficient training and improved performance in generative modeling tasks.

Quantum Annealing for Machine Learning

Quantum annealing is a quantum-native approach that solves optimization problems by finding the ground state of a quantum system. This method has been successfully applied to machine learning tasks such as clustering, graph-based learning, and feature selection. Companies like D-Wave Systems have demonstrated the utility of quantum annealing in solving real-world problems, paving the way for broader adoption of quantum-native methods.

Challenges and Future Directions

While QML holds immense promise, several challenges must be addressed to realize its full potential. Quantum hardware is still in its infancy, with issues such as limited qubit coherence, noise, and scalability posing significant barriers. Additionally, the development of quantum algorithms and the integration of quantum systems into existing AI frameworks require interdisciplinary expertise and continued research.

Looking ahead, advancements in fault-tolerant quantum computing, error correction, and hybrid architectures will play a crucial role in overcoming these challenges. As the field matures, the potential for QML to revolutionize industries ranging from healthcare to finance becomes increasingly tangible.

Conclusion

Quantum machine learning represents a frontier of innovation, bridging the gap between quantum computing and AI. Through hybrid quantum-classical approaches and quantum-native algorithms, QML is poised to redefine how we process data, optimize models, and solve complex problems. Drawing on the foundational work of Biamonte et al. (2017) and Schuld and Petruccione (2018), this chapter has highlighted the transformative potential of QML in shaping the future of AI.

In the next chapter, we will explore how AI can, in turn, enhance quantum computing systems, focusing on the role of AI in quantum hardware design and error correction.

AI MEETS QUANTUM

NAJEED KHAN

CHAPTER 4: AI FOR QUANTUM SYSTEM DESIGN

The development of quantum computing hardware and systems is fraught with challenges, ranging from designing optimal quantum circuits to addressing error correction in noisy intermediate-scale quantum (NISQ) devices. Artificial intelligence (AI) offers a transformative approach to overcome these hurdles by automating design processes and optimizing performance. In this chapter, we explore how AI accelerates quantum system design and enhances quantum error correction, drawing on insights from Krenn et al. (2020) and Carleo et al. (2017).

AI-Driven Quantum Hardware Design

Quantum hardware design involves creating quantum circuits and architectures capable of executing computations with high fidelity and efficiency. Traditional methods for designing quantum systems rely on human intuition and manual experimentation, which are often time-consuming and prone to error. AI introduces a paradigm shift by automating these processes and discovering novel designs beyond human capability.

Automated Quantum Experiments

Krenn et al. (2020) demonstrated the potential of AI to automate quantum experiments. By using machine learning algorithms, researchers developed systems capable of generating and optimizing quantum optical experiments. These algorithms explore vast design spaces, identifying configurations that maximize the performance of quantum systems.

For example, AI can design quantum circuits with specific functionalities, such as generating entangled states or implementing quantum gates. By simulating millions of configurations and iteratively refining them, AI-driven tools significantly reduce the time required to develop and test quantum hardware.

Enhancing Qubit Connectivity and Scaling

Qubit connectivity is a critical factor in the performance of quantum systems. AI aids in optimizing qubit layouts and connectivity graphs to minimize errors and maximize computational efficiency. Neural network models can predict the performance of various configurations and suggest modifications that improve scalability and robustness.

Optimizing Quantum Error Correction

Quantum error correction (QEC) is essential for mitigating the effects of noise and decoherence in quantum systems. Traditional QEC methods, such as surface codes, involve complex error-detection and correction procedures. AI enhances these processes by identifying optimal error-correction strategies and improving their implementation.

Neural Networks for Error Detection

Carleo et al. (2017) introduced the use of artificial neural networks to solve quantum many-body problems. Building on this approach, neural networks have been applied to QEC tasks, such as error detection and syndrome decoding. By training on simulated error patterns, these networks learn to identify and correct errors with high accuracy, even in the presence of noisy data.

Adaptive Error Mitigation

AI-driven adaptive algorithms dynamically adjust error-correction strategies based on the real-time state of the quantum system. For instance, reinforcement learning techniques enable AI agents to learn optimal policies for applying error-correction codes. These adaptive methods improve the resilience of quantum systems to noise and extend the coherence times of qubits.

AI for Quantum Many-Body Systems

The design and optimization of quantum many-body systems —complex systems of interacting quantum particles—present unique challenges. Carleo et al. (2017) demonstrated how neural networks, specifically restricted Boltzmann machines (RBMs), can represent quantum states and solve many-body problems. This approach has profound implications for quantum hardware design, enabling the simulation and analysis of intricate quantum systems.

By leveraging AI to model quantum many-body interactions,

researchers gain insights into the behavior of quantum systems under various conditions. These insights inform the design of hardware architectures optimized for specific applications, such as quantum simulation and material science.

Challenges and Future Directions

Despite its potential, the application of AI to quantum system design is not without challenges. The complexity of quantum mechanics and the limitations of current AI models require continued innovation and interdisciplinary collaboration. Additionally, the integration of AI with quantum hardware necessitates robust interfaces and scalable frameworks.

Future advancements in AI algorithms, combined with improvements in quantum hardware, will unlock new possibilities for quantum system design. From automating the discovery of novel quantum phenomena to refining error-correction techniques, the synergy between AI and quantum computing is poised to accelerate the realization of practical quantum technologies.

Conclusion

AI plays a pivotal role in advancing quantum system design, from automating hardware development to optimizing error correction. By leveraging the capabilities of machine learning and neural networks, researchers are addressing key challenges in quantum computing and paving the way for scalable, high-performance quantum systems. Building on the foundational work of Krenn et al. (2020) and Carleo et al. (2017), this chapter underscores the transformative potential of AI in shaping the future of quantum technologies.

In the next chapter, we will explore the applications of AI and quantum computing in the natural sciences, focusing on their synergistic impact in fields such as chemistry, biology, and physics.

NAJEED KHAN

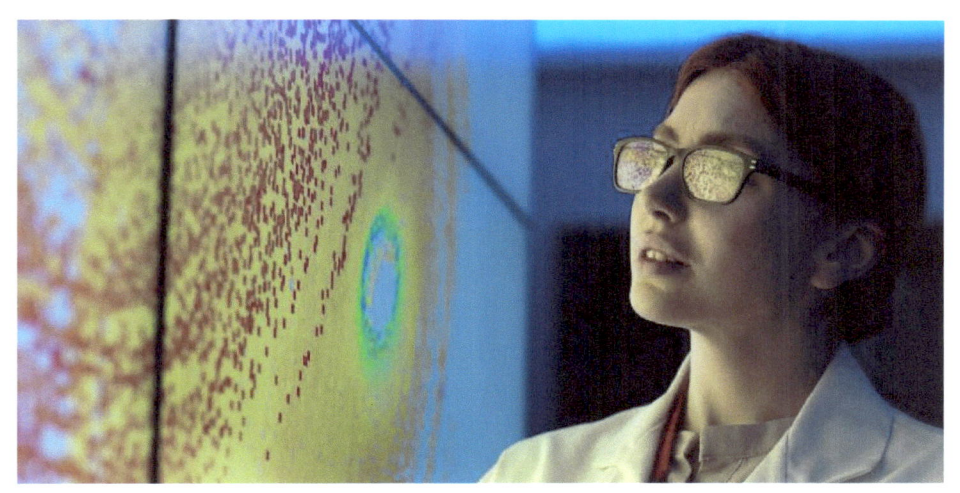

CHAPTER 5: APPLICATIONS IN NATURAL SCIENCES

The natural sciences—chemistry, biology, and physics—stand to benefit profoundly from the convergence of artificial intelligence (AI) and quantum computing. This synergy offers unprecedented capabilities for modeling complex systems, solving intractable problems, and discovering new phenomena. In this chapter, we explore how AI and quantum computing combine to advance the frontiers of these disciplines, drawing from insights provided by Cao et al. (2019) and Aspuru-Guzik et al. (2018).

Quantum Chemistry: Transforming Molecular Simulations

Quantum chemistry is one of the most promising areas for quantum computing applications. Accurate modeling of molecular systems often requires solving the Schrödinger equation for many-body quantum systems, a task that scales exponentially with system size and quickly becomes computationally infeasible for classical methods. Quantum computing, augmented by AI, offers a transformative solution.

Solving Molecular Systems with Quantum Computing

As highlighted by Cao et al. (2019), quantum computers can efficiently simulate molecular systems by directly encoding quantum states and operations. Algorithms like the Variational

Quantum Eigensolver (VQE) and Quantum Phase Estimation (QPE) allow researchers to calculate molecular ground states and reaction dynamics. AI enhances these simulations by optimizing quantum circuits and parameter tuning, improving both accuracy and efficiency.

Accelerating Drug Discovery and Material Design

By combining quantum computing's ability to simulate complex molecules with AI's pattern recognition capabilities, researchers can accelerate drug discovery and material design. AI algorithms identify promising candidate molecules, which are then analyzed using quantum simulations to predict their properties with unprecedented precision. This iterative approach reduces the time and cost associated with experimental trials, enabling the rapid development of new pharmaceuticals and advanced materials.

Biology: Decoding Complex Biological Systems

The intricate mechanisms of biological systems present a significant challenge for computational modeling. Quantum computing, integrated with AI, provides tools for unraveling these complexities, offering insights into protein folding, genetic interactions, and cellular processes.

Protein Folding and Genomic Analysis

Protein folding, a problem of immense biological and medical significance, involves predicting the three-dimensional structure of a protein from its amino acid sequence. Quantum computing can simulate the quantum interactions governing protein folding, while AI assists in analyzing the vast data generated by these simulations. This combination has the potential to unlock new therapies for diseases linked to misfolded proteins.

In genomic analysis, AI identifies patterns in genetic data, while quantum algorithms enable efficient processing of large genomic datasets. Together, these technologies facilitate advancements in personalized medicine, gene editing, and understanding the genetic basis of diseases.

Physics: Unlocking Fundamental Insights

Physics has long been a driving force behind both AI and quantum computing. The integration of these technologies enables breakthroughs in understanding complex physical phenomena, from condensed matter systems to quantum field theory.

Simulation of Condensed Matter Systems

Condensed matter physics studies the behavior of many-body systems, such as superconductors and quantum magnets. As noted by Aspuru-Guzik et al. (2018), quantum computing excels at simulating these systems, providing insights into their emergent properties. AI further aids in interpreting simulation results and identifying new phases of matter.

Advancing Quantum Field Theory

Quantum field theory (QFT), which underpins much of modern physics, involves calculations that are computationally prohibitive for classical methods. Quantum computers can perform QFT calculations more efficiently, while AI algorithms assist in identifying patterns and optimizing computational workflows. This synergy has the potential to deepen our understanding of fundamental forces and particles.

Challenges and Future Directions

While the applications of AI and quantum computing in the natural sciences are vast, several challenges remain. Quantum hardware limitations, such as qubit coherence and error rates, must be addressed to realize the full potential of these technologies. Similarly, the integration of AI with quantum systems requires advancements in algorithms, data preprocessing, and interdisciplinary collaboration.

Future research will likely focus on hybrid quantum-classical systems that combine the strengths of both paradigms. As these technologies mature, their impact on the natural sciences will continue to grow, driving discoveries that were once thought to be out of reach.

Conclusion

The intersection of AI and quantum computing is revolutionizing the natural sciences, enabling breakthroughs in chemistry, biology, and physics. By leveraging the unique strengths of these technologies, researchers are solving problems of unprecedented complexity and advancing our understanding of the natural world. Drawing from the foundational work of Cao et al. (2019) and Aspuru-Guzik et al. (2018), this chapter has highlighted the transformative potential of AI-quantum synergies in the natural sciences.

In the next chapter, we will examine the implications of these technologies for cryptography, exploring how AI and quantum computing are reshaping data security and encryption.

NAJEED KHAN

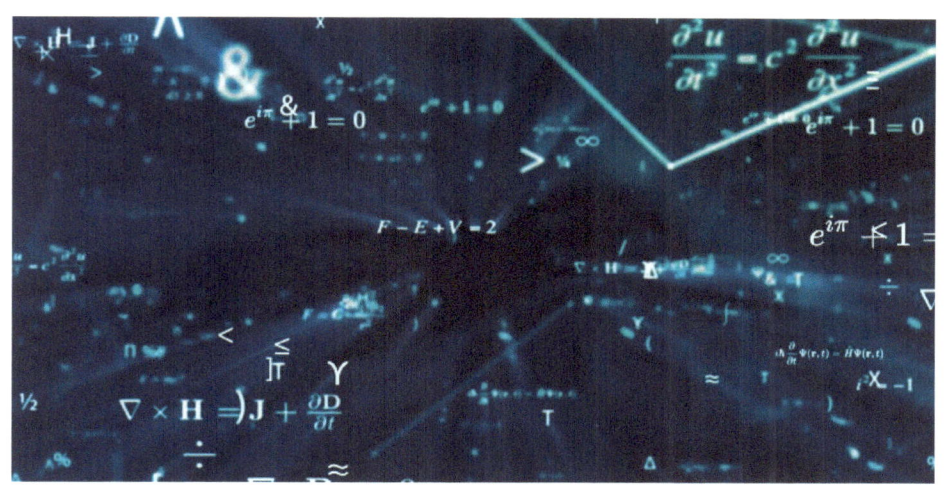

CHAPTER 6: REVOLUTIONIZING CRYPTOGRAPHY

Cryptography underpins the security of modern digital communications, safeguarding sensitive data against unauthorized access. However, the convergence of artificial intelligence (AI) and quantum computing is poised to disrupt this landscape. While quantum computing introduces new vulnerabilities to traditional encryption schemes, it also offers innovative tools to enhance cryptographic techniques. In this chapter, we explore the dual role of AI and quantum computing in revolutionizing cryptography, leveraging foundational insights from Shor (1997) and Bernstein et al. (2009).

Quantum Computing's Threat to Classical Cryptography

Traditional cryptographic systems rely on the computational difficulty of certain mathematical problems, such as integer factorization and discrete logarithms. For instance, RSA encryption depends on the infeasibility of factoring large integers, while elliptic curve cryptography (ECC) relies on the hardness of solving discrete logarithms over elliptic curves. These problems are computationally secure against classical computers but are vulnerable to quantum algorithms.

Shor's Algorithm

As demonstrated by Shor (1997), quantum computers can solve

integer factorization and discrete logarithms in polynomial time. Shor's algorithm leverages quantum parallelism and interference to efficiently find the prime factors of large numbers. This capability renders widely used encryption schemes, such as RSA and ECC, obsolete in a post-quantum era. The advent of practical quantum computers would thus undermine the security of digital communications, e-commerce, and sensitive government data.

Post-Quantum Cryptography

In response to the threats posed by quantum computing, researchers are developing post-quantum cryptography (PQC) schemes that remain secure against quantum attacks. As discussed by Bernstein et al. (2009), PQC focuses on cryptographic methods based on problems that are resistant to both classical and quantum algorithms. Examples include lattice-based cryptography, code-based cryptography, and multivariate polynomial cryptography.

Role of AI in PQC

AI plays a significant role in advancing PQC by:

- **Designing Robust Schemes:** Machine learning algorithms assist in identifying weaknesses in proposed cryptographic schemes by simulating potential attack vectors.
- **Optimizing Implementations:** AI optimizes the performance of PQC algorithms for practical deployment, ensuring efficient use of computational resources.

- **Automating Security Analysis:** AI systems analyze cryptographic protocols to detect vulnerabilities and recommend improvements, enhancing their resistance to quantum attacks.

AI-Enhanced Cryptographic Techniques

Beyond defending against quantum threats, AI contributes to the evolution of cryptography through the development of advanced techniques for data security.

Ai-Driven Key Management

Key management is a critical aspect of cryptographic systems. AI algorithms improve key generation, distribution, and storage by identifying patterns in usage and optimizing processes. For example, neural networks can predict potential vulnerabilities in key management workflows and suggest mitigations to enhance security.

Cryptographic Protocol Design

AI aids in designing complex cryptographic protocols by automating the discovery of optimal configurations. These protocols incorporate advanced techniques, such as zero-knowledge proofs and homomorphic encryption, which enable secure computation and data sharing without compromising privacy.

Quantum Cryptography: Harnessing

Quantum Properties for Security

While quantum computing poses a threat to classical cryptography, it also enables the development of fundamentally secure cryptographic methods based on quantum mechanics. Quantum key distribution (QKD) is a prominent example.

Quantum Key Distribution

QKD protocols, such as BB84, use quantum properties like superposition and entanglement to establish cryptographic keys. The laws of quantum mechanics ensure that any eavesdropping attempt introduces detectable anomalies, guaranteeing secure communication channels. AI enhances QKD by optimizing error correction and signal processing, improving the reliability and scalability of quantum communication networks.

Quantum Random Number Generation

Randomness is a cornerstone of secure cryptography. Quantum random number generators (QRNGs) leverage the inherent unpredictability of quantum phenomena to produce true random numbers. AI algorithms analyze and validate the quality of randomness, ensuring it meets cryptographic standards.

Challenges and Future Directions

The integration of AI and quantum computing into cryptography is not without challenges. Key issues include:

- **Scalability:** Developing scalable PQC and quantum cryptographic solutions suitable for global adoption.

- **Interoperability:** Ensuring compatibility between classical, quantum, and AI-enhanced cryptographic systems.
- **Trust and Transparency:** Building trust in AI-driven cryptographic methods by ensuring transparency in algorithm design and implementation.

Future research will focus on bridging the gap between theoretical advancements and practical deployment. The collaboration of cryptographers, quantum physicists, and AI researchers will be essential in addressing these challenges.

Conclusion

The convergence of AI and quantum computing represents both a challenge and an opportunity for cryptography. While quantum computing threatens the foundations of traditional encryption, it also enables the development of revolutionary cryptographic techniques. AI, in turn, enhances the design, analysis, and implementation of secure systems, ensuring resilience against emerging threats. Drawing on the pioneering work of Shor (1997) and Bernstein et al. (2009), this chapter highlights the transformative potential of AI and quantum computing in shaping the future of cryptographic security.

In the next chapter, we will explore how the fusion of AI and quantum computing is driving innovation in the financial sector, with applications in risk analysis, portfolio optimization, and fraud detection.

NAJEED KHAN

CHAPTER 7: QUANTUM-AI FUSION IN FINANCE

The financial sector is a natural arena for the convergence of artificial intelligence (AI) and quantum computing. Both technologies bring transformative potential to financial operations, enabling advanced solutions for risk analysis, portfolio optimization, and fraud detection. By leveraging the insights of Orús et al. (2019) and Egger et al. (2020), this chapter explores how quantum-AI systems are revolutionizing finance, addressing both current applications and future possibilities.

Risk Analysis: Enhancing Predictive Models

Risk management is at the core of financial decision-making. Institutions must evaluate potential losses under various market scenarios, a task that involves analyzing vast amounts of data and modeling complex interactions.

AI-Powered Risk Assessment

AI has already revolutionized risk analysis by enabling real-time data processing and predictive modeling. Machine learning algorithms analyze historical data to predict market movements, assess creditworthiness, and identify emerging risks. However, the computational demands of these models grow with data complexity and scale.

Quantum Computing for Scenario Simulation

Quantum computing provides a complementary approach by accelerating scenario simulations. Quantum algorithms, such as the Quantum Monte Carlo method, allow for more efficient sampling of probability distributions compared to classical counterparts. As noted by Orús et al. (2019), quantum systems excel at solving stochastic processes that underpin financial risk models, enabling more accurate and granular assessments.

Hybrid Quantum-AI Models

By combining AI's data-driven insights with quantum computing's simulation capabilities, hybrid systems can enhance risk analysis. For instance, AI identifies patterns and trends in market data, while quantum algorithms simulate potential outcomes under various economic conditions. This synergy results in faster and more reliable risk evaluations, empowering financial institutions to make informed decisions.

Portfolio Optimization: Balancing Risk and Return

Portfolio optimization aims to allocate assets in a way that maximizes returns while minimizing risks. This task involves solving optimization problems that grow exponentially in complexity with the number of assets.

Quantum Algorithms for Optimization

Quantum computing introduces algorithms like the Quantum Approximate Optimization Algorithm (QAOA) to solve complex optimization problems. As highlighted by Egger et al. (2020), QAOA can efficiently explore large solution spaces, finding near-optimal asset allocations that balance risk and return. This approach is particularly valuable for high-dimensional portfolios where classical methods struggle.

AI in Dynamic Portfolio Management

AI enhances portfolio optimization by adapting to changing market conditions. Machine learning models analyze real-time data to adjust asset allocations dynamically, ensuring portfolios remain aligned with investors' objectives. When integrated with quantum optimization, AI systems can test multiple scenarios and strategies simultaneously, providing more robust recommendations.

Fraud Detection: Combating Financial Crime

Fraud detection is another critical application where quantum-AI fusion demonstrates significant potential. Detecting fraudulent activities requires analyzing vast transactional datasets to identify anomalies and suspicious patterns.

AI-Driven Anomaly Detection

AI excels at detecting fraud through techniques such as anomaly detection and deep learning. These methods identify deviations from normal behavior, flagging potentially fraudulent transactions. However, the scalability of AI models is limited by the computational demands of processing large datasets in real-time.

Quantum Speedup for Pattern Recognition

Quantum computing can accelerate pattern recognition tasks by leveraging quantum-enhanced machine learning algorithms. As described by Egger et al. (2020), quantum systems process high-dimensional data more efficiently, enabling faster detection of fraud. For example, quantum support vector machines and quantum neural networks can classify transactions with greater speed and accuracy than their classical counterparts.

Challenges and Future Directions

While the fusion of AI and quantum computing holds immense promise for finance, several challenges must be addressed:

- **Hardware Limitations:** Current quantum computers are constrained by qubit coherence and error rates, limiting their practical applications.
- **Algorithm Development:** Designing effective quantum

algorithms tailored for financial use cases remains an active area of research.

- **Integration:** Seamlessly integrating AI and quantum systems into existing financial infrastructure requires overcoming technical and operational hurdles.

Future developments will likely focus on advancing quantum hardware, creating robust hybrid architectures, and fostering collaboration between academia and industry. As these challenges are addressed, quantum-AI systems will play an increasingly central role in financial innovation.

Conclusion

The integration of AI and quantum computing is poised to revolutionize the financial sector, transforming how institutions manage risk, optimize portfolios, and combat fraud. By leveraging the insights of Orús et al. (2019) and Egger et al. (2020), this chapter has outlined the transformative potential of quantum-AI fusion in finance. As these technologies mature, they will enable financial institutions to navigate an increasingly complex and dynamic global economy with greater efficiency and confidence.

In the next chapter, we will examine how AI and quantum computing are driving breakthroughs in healthcare and drug discovery, accelerating the development of personalized medicine and novel therapies.

NAJEED KHAN

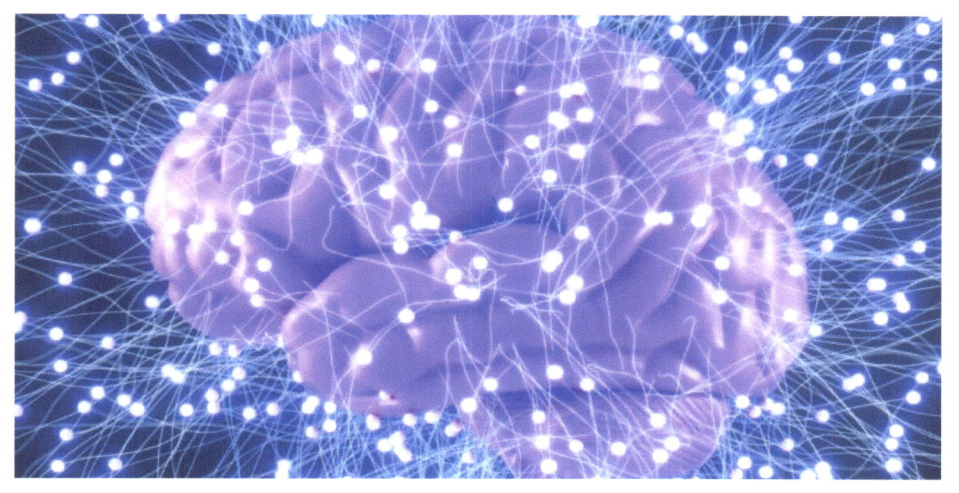

CHAPTER 8: HEALTHCARE AND DRUG DISCOVERY

The convergence of artificial intelligence (AI) and quantum computing is poised to revolutionize healthcare and drug discovery. These cutting-edge technologies promise breakthroughs in genomic analysis, drug development, and personalized medicine, addressing some of the most challenging problems in the medical domain. Drawing on insights from Topol (2019) and Perdomo-Ortiz et al. (2012), this chapter examines how quantum-AI technologies are accelerating innovation and improving patient outcomes.

Accelerating Drug Discovery

Drug discovery is a resource-intensive process, requiring years of research and billions of dollars in investment. Traditional methods involve extensive computational simulations and experimental validation, which are often limited by the complexity of molecular interactions.

Quantum Computing in Molecular Modeling

Quantum computing excels at simulating quantum mechanical systems, such as molecular interactions. Unlike classical computers, which approximate these interactions, quantum computers can represent them with higher fidelity. As demonstrated by Perdomo-Ortiz et al. (2012), quantum annealing can identify low-energy conformations of protein models, enabling more accurate predictions of molecular behavior.

By leveraging quantum algorithms, researchers can:

- **Optimize Lead Compounds:** Identify promising drug candidates by exploring vast chemical spaces more efficiently.

- **Simulate Complex Interactions:** Accurately model the binding of drugs to target proteins, reducing the need for trial-and-error experiments.

- **Predict Side Effects:** Anticipate adverse effects by analyzing interactions with off-target proteins.

AI-Driven Drug Discovery

AI complements quantum computing by analyzing large datasets and predicting outcomes. Machine learning models identify patterns in biological data, accelerating target identification and compound screening. For example, AI can:

- **Analyze Omics Data:** Process genomic, transcriptomic, and proteomic data to identify disease pathways.

- **Prioritize Candidates:** Rank drug candidates based on predicted efficacy and safety profiles.

- **Automate Experiments:** Optimize laboratory workflows through robotics and AI-driven decision-making.

Genomic Analysis

Genomic data analysis is essential for understanding the genetic basis of diseases and tailoring treatments to individual patients. However, the complexity of genomic datasets poses significant computational challenges.

Quantum Algorithms for Genomic Sequencing

Quantum algorithms enable faster and more accurate analysis of genomic data. By solving combinatorial optimization problems inherent in sequence alignment and variant calling, quantum computing can:

- **Improve Accuracy:** Enhance the detection of genetic variants associated with diseases.
- **Reduce Computational Costs:** Perform large-scale analyses more efficiently than classical methods.

AI in Genomics

AI has already transformed genomics through deep learning and natural language processing. These techniques extract insights from complex datasets, such as gene expression profiles and epigenetic markers. When combined with quantum computing, AI can:

- **Uncover Hidden Patterns:** Identify rare genetic mutations linked to specific conditions.
- **Predict Gene-Environment Interactions:** Model how environmental factors influence genetic predispositions.
- **Enable Real-Time Diagnostics:** Analyze genomic data at

the point of care, supporting precision medicine.

Personalized Medicine

Personalized medicine aims to tailor treatments to individual patients based on their unique genetic, environmental, and lifestyle factors. Quantum-AI systems enhance this approach by enabling more accurate diagnostics and treatment recommendations.

AI for Patient Stratification

AI algorithms stratify patients into subgroups based on their genetic and clinical profiles. This stratification informs treatment decisions, ensuring that patients receive therapies most likely to benefit them.

Quantum-Enhanced Drug Matching

Quantum computing accelerates the identification of optimal drug combinations for individual patients. By simulating the interactions between multiple drugs and patient-specific molecular profiles, quantum algorithms can:

- **Optimize Dosing Regimens:** Recommend precise dosages to maximize efficacy and minimize side effects.
- **Identify Synergistic Therapies:** Discover drug combinations that work better together than individually.

Challenges and Ethical Considerations

The integration of quantum-AI technologies in healthcare raises several challenges and ethical concerns:

- **Data Privacy:** Ensuring the security of sensitive patient data during analysis and storage.
- **Equity:** Addressing disparities in access to advanced technologies, particularly in underserved communities.
- **Validation:** Establishing rigorous standards for validating quantum-AI models in clinical settings.

Topol (2019) emphasizes the importance of maintaining a human-centered approach to healthcare, ensuring that technological advancements enhance, rather than replace, the physician-patient relationship.

Future Directions

The future of quantum-AI in healthcare lies in:

- **Scalable Infrastructure:** Developing quantum hardware and AI systems capable of handling real-world medical datasets.
- **Collaborative Research:** Fostering partnerships between researchers, clinicians, and industry stakeholders.
- **Regulatory Frameworks:** Establishing guidelines for the safe and ethical deployment of these technologies.

Conclusion

Quantum computing and AI are set to transform healthcare by accelerating drug discovery, advancing genomic analysis, and enabling personalized medicine. By harnessing the power of these technologies, researchers and clinicians can tackle complex medical challenges, ultimately improving patient outcomes and reducing healthcare costs. Drawing on foundational work by Topol (2019) and Perdomo-Ortiz et al. (2012), this chapter highlights the potential of quantum-AI systems to drive innovation and reshape the future of medicine.

In the next chapter, we will explore how AI and quantum computing are advancing quantum networking and building secure communication systems for the future.

NAJEED KHAN

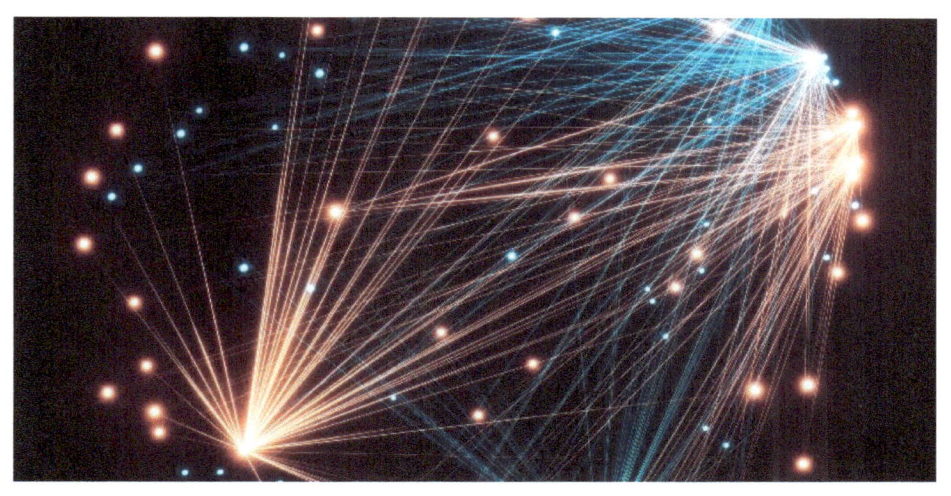

CHAPTER 9: AI-ENHANCED QUANTUM NETWORKING

Quantum networking represents the next frontier in communication technology, enabling the secure exchange of quantum information and laying the groundwork for the quantum internet. The integration of artificial intelligence (AI) into quantum networks has accelerated progress toward scalable, efficient, and secure systems. Building on the foundational work of Kimble (2008) and Van Meter and Devitt (2016), this chapter explores how AI enhances quantum networking and advances the development of a global quantum communication infrastructure.

The Vision of the Quantum Internet

Kimble (2008) conceptualized the quantum internet as a network capable of transmitting quantum information using quantum entanglement and superposition. Such a network promises transformative applications, including:

- **Unbreakable Encryption:** Quantum key distribution (QKD) ensures communication security against eavesdropping.
- **Distributed Quantum Computing:** Connecting quantum computers to solve problems beyond the reach of classical systems.
- **Precision Sensing:** Enabling collaborative quantum sensors for scientific and industrial applications.

Challenges in Quantum Networking

Despite its potential, quantum networking faces significant technical challenges:

- **Fragility of Quantum States:** Quantum information is susceptible to decoherence and loss during transmission.
- **Scalability:** Creating large-scale networks with interconnected nodes requires overcoming physical and logistical barriers.
- **Error Correction:** Ensuring reliable communication necessitates robust quantum error correction mechanisms.

AI's Role in Advancing Quantum Networking

AI plays a critical role in addressing the challenges of quantum networking by optimizing network design, improving error correction, and enabling dynamic resource allocation.

Optimizing Network Topology

AI algorithms analyze network configurations to optimize the placement of quantum nodes and links. By simulating various topologies, AI can:

- **Minimize Latency:** Ensure efficient routing of quantum information.
- **Maximize Entanglement Distribution:** Enhance the

availability of entangled qubits across the network.

- **Adapt to Scalability:** Design networks that grow seamlessly as more nodes are added.

Enhancing Quantum Error Correction

Error correction is vital for maintaining the integrity of quantum information during transmission. Traditional methods are computationally expensive and scale poorly with network size. AI provides a more efficient approach:

- **Error Pattern Recognition:** Machine learning models detect and classify error patterns in real-time.
- **Adaptive Correction Protocols:** AI systems dynamically adjust error correction strategies based on network conditions.
- **Resource Optimization:** Reduce the computational overhead of error correction, freeing up resources for other tasks.

Dynamic Resource Management

Quantum networks require dynamic allocation of resources, such as qubits, entanglement links, and bandwidth. AI enhances resource management by:

- **Predictive Analytics:** Forecasting network demand to allocate resources proactively.
- **Real-Time Optimization:** Adjusting resource allocation to meet changing requirements.
- **Fault Tolerance:** Identifying and rerouting around failed nodes or links to maintain network integrity.

Building Secure Quantum Networks

Security is a cornerstone of quantum networking, with AI playing a pivotal role in strengthening network defenses.

Quantum Key Distribution (Qkd)

Quantum key distribution enables secure communication by leveraging the principles of quantum mechanics. AI enhances QKD by:

- **Optimizing Protocols:** Refining QKD algorithms to maximize key generation rates and minimize losses.
- **Intrusion Detection:** Using AI to identify and mitigate potential eavesdropping attempts.

Ai-Driven Security Monitoring

AI systems continuously monitor network activity, detecting anomalies and potential threats. By combining classical security measures with quantum principles, AI creates a multilayered defense strategy that:

- **Identifies Vulnerabilities:** Proactively addresses weaknesses in the network.
- **Responds to Attacks:** Implements countermeasures in real-time to secure communication.

Toward Scalable Distributed Quantum Computing

Van Meter and Devitt (2016) emphasized the importance of scalability in distributed quantum computing. AI facilitates this goal by:

- **Node Synchronization:** Coordinating operations across geographically dispersed quantum nodes.
- **Workflow Optimization:** Managing computational tasks to maximize network efficiency.
- **Interoperability:** Ensuring compatibility between different quantum technologies and platforms.

Future Directions and Innovations

The integration of AI into quantum networking continues to drive innovation in the following areas:

- **Hybrid Quantum-Classical Networks:** Combining classical and quantum resources to create versatile communication systems.
- **Intelligent Quantum Routers:** AI-powered devices that dynamically manage entanglement distribution and routing.
- **Cross-Disciplinary Collaboration:** Bridging expertise from AI, quantum computing, and telecommunications to accelerate progress.

Conclusion

AI-enhanced quantum networking represents a transformative step toward realizing the vision of a global quantum internet.

By addressing challenges in scalability, error correction, and resource management, AI enables the efficient and secure exchange of quantum information. Building on the foundational ideas of Kimble (2008) and Van Meter and Devitt (2016), this chapter highlights the critical role of AI in advancing quantum communications. As research and development continue, AI-driven quantum networks will unlock new possibilities for science, industry, and society.

The final chapter will explore the ethical considerations and challenges at the intersection of AI and quantum technologies, ensuring these advancements align with societal values and priorities.

NAJEED KHAN

CHAPTER 10: ETHICS AND CHALLENGES IN QUANTUM-AI SYSTEMS

The intersection of artificial intelligence (AI) and quantum computing has the potential to reshape industries, economies, and even societal structures. However, as with any transformative technology, this convergence raises profound ethical questions, societal implications, and technical challenges. Drawing on the insights of Tegmark (2017) and Bostrom (2014), this chapter explores these critical considerations to ensure that the advancements in quantum-AI systems align with humanity's values and priorities.

Ethical Concerns at the Quantum-AI Interface

Bias And Fairness

The integration of AI and quantum technologies could amplify existing biases embedded in datasets and algorithms. Key concerns include:

- **Quantum-AI Bias:** Ensuring that quantum-enhanced AI systems do not perpetuate or exacerbate inequities.
- **Access Disparities:** Preventing the monopolization of quantum-AI technologies by a few powerful entities, which could deepen global inequality.
- **Accountability:** Establishing clear frameworks for responsibility when decisions made by quantum-AI systems have significant consequences.

Data Privacy

Quantum-AI systems are poised to revolutionize data processing capabilities, but this raises significant privacy concerns:

- **Decryption Capabilities:** Quantum computing's potential to break classical cryptographic protocols endangers the security of sensitive information.
- **AI-Driven Surveillance:** The combination of quantum computing and AI could enable unprecedented surveillance capabilities, necessitating robust oversight mechanisms.

Autonomy And Control

As Bostrom (2014) highlighted, advanced AI systems—especially those integrated with quantum technologies—pose risks of losing human control over decision-making processes. Critical questions include:

- **Ethical Autonomy:** How much decision-making autonomy should be entrusted to quantum-AI systems?
- **Human Oversight:** Ensuring that human judgment remains central in critical domains, such as healthcare and governance.

Societal Impacts

Economic Disruption

Quantum-AI technologies promise to unlock significant efficiencies but may also disrupt labor markets and economic systems:

- **Automation of Complex Tasks:** Jobs requiring advanced problem-solving or data analysis could become obsolete.
- **New Industries:** The rise of quantum-AI industries will create opportunities but may leave behind those without access to necessary education and resources.
- **Wealth Concentration:** The development and deployment of these technologies by a select few entities could lead to greater economic disparities.

National Security

Quantum-AI systems have profound implications for national security:

- **Arms Race:** Nations may compete to dominate quantum-AI capabilities, leading to a new kind of technological arms race.
- **Cybersecurity Threats:** The ability to crack existing cryptographic systems could destabilize global communications and financial systems.
- **Global Cooperation:** Balancing competition with the need for international collaboration on ethical and security standards will be critical.

Environmental Considerations

The computational power of quantum-AI systems comes with significant energy requirements:

- **Sustainable Development:** Ensuring that the energy demands of quantum computing and AI align with global climate goals.

- **Resource Management:** Addressing the environmental costs of producing quantum hardware and training AI models.

Technical Challenges

Scalability And Error Correction

Building scalable quantum-AI systems requires overcoming significant technical hurdles:

- **Quantum Decoherence:** Maintaining the stability of quantum states over extended periods remains a fundamental challenge.
- **Error Rates:** Quantum systems are inherently error-prone, and integrating them with AI requires robust error correction techniques.

Interoperability

Integrating classical, quantum, and AI systems seamlessly is a non-trivial task:

- **Hybrid Architectures:** Designing systems that leverage the strengths of both quantum and classical computing.
- **Standardization:** Developing universal protocols and standards for quantum-AI integration.

Ethical Algorithms

Embedding ethical considerations into quantum-AI systems is a complex challenge:

- **Value Alignment:** Ensuring that these systems operate in alignment with human values.
- **Explainability:** Making the decision-making processes of quantum-AI systems transparent and understandable.

Navigating the Path Forward

Multi-Stakeholder Collaboration

Tegmark (2017) and Bostrom (2014) emphasize the importance of involving diverse stakeholders in shaping the future of transformative technologies:

- **Policymakers:** Establishing regulations that balance innovation with ethical safeguards.
- **Academia:** Conducting interdisciplinary research to address technical, ethical, and societal challenges.
- **Industry Leaders:** Promoting responsible innovation and equitable access to quantum-AI technologies.
- **Civil Society:** Engaging the public in discussions about the implications of these technologies.

Ethical Frameworks

Developing comprehensive ethical frameworks is essential to guide the development and deployment of quantum-AI systems:

- **Principles for Responsible AI:** Extending existing AI ethics principles to encompass quantum technologies.
- **International Guidelines:** Establishing global norms and agreements to prevent misuse and ensure equitable

benefits.

Education And Awareness

Raising awareness about the implications of quantum-AI technologies is crucial:

- **Public Engagement:** Educating society about the opportunities and risks associated with these technologies.
- **Workforce Development:** Preparing individuals for jobs in emerging quantum-AI industries through education and training programs.

Conclusion

The convergence of AI and quantum computing holds immense promise but also demands thoughtful consideration of its ethical, societal, and technical dimensions. By addressing these challenges proactively, we can harness the power of quantum-AI systems to drive progress while safeguarding humanity's values and priorities. As Tegmark (2017) and Bostrom (2014) argue, the choices we make today will shape the trajectory of these transformative technologies for generations to come.

This chapter concludes the exploration of the synergies and future applications of AI and quantum computing. As we stand at the cusp of this new era, it is imperative to navigate its challenges with wisdom, collaboration, and a steadfast commitment to ethical progress.

NAJEED KHAN

BIBLIOGRAPHY

1. Bernstein, D. J., et al. (2009). *Post-Quantum Cryptography*. Springer.
2. Biamonte, J., et al. (2017). Quantum machine learning. *Nature*, *549*(7671), 195-202.
3. Carleo, G., et al. (2017). Solving the quantum many-body problem with artificial neural networks. *Science*, *355*(6325).
4. Cao, Y., et al. (2019). Quantum chemistry in the age of quantum computing. *Chemical Reviews*, *119*(19), 10856-10915.
5. Farhi, E., et al. (2014). A quantum approximate optimization algorithm. *arXiv preprint arXiv:1411.4028*.
6. Kimble, H. J. (2008). The quantum internet. *Nature*, *453*(7198), 1023-1030.
7. Nielsen, M. A., & Chuang, I. L. (2010). *Quantum Computation and Quantum Information*. Cambridge University Press.
8. Schuld, M., & Petruccione, F. (2018). *Supervised Learning with Quantum Computers*. Springer.
9. Schuld, M., & Killoran, N. (2019). Quantum machine learning in feature Hilbert spaces. *Physical Review Letters*, *122*(4).
10. Tegmark, M. (2017). *Life 3.0: Being Human in the Age of Artificial Intelligence*. Knopf.
11. Topol, E. J. (2019). *Deep Medicine: How Artificial Intelligence Can Make Healthcare Human Again*. Basic Books.

12. Van Meter, R., & Devitt, S. J. (2016). The path to scalable distributed quantum computing. *Computer*, *49*(9), 31-42.

AI MEETS QUANTUM

ABOUT THE AUTHOR

Najeed Khan

Najeed Khan is a strategic business leader and technology expert with over two decades of experience shaping the intersection of business strategy and emerging technologies. As a trusted advisor to C-suite executives at leading technology companies, he has pioneered approaches that bridge the gap between technical innovation and business value.

A recognized thought leader in the autonomous economy, Khan frequently contributes to industry discussions and has served as a panel speaker at prestigious technology forums. His insights are shaped by years of experience across global organizations.

Khan holds particular expertise in strategy, with a proven track record of translating complex technical concepts into actionable business strategies. His approach combines rigorous analytical thinking with practical business acumen, making him a valued voice in discussions about technology's role in shaping enterprise strategy.

Khan holds an MBA from NYU Stern School of Business and a B.E. (Honors) in Electrical and Electronics Engineering from Birla Institute of Technology and Science, Pilani. The combination of technical expertise and business experience has shaped his unique approach to technology strategy, making him a distinctive voice in discussions about the future of enterprise technology and digital transformation.

BOOKS BY THIS AUTHOR

Ai Demystified: The Fundamentals And Beyond

https://www.amazon.com/dp/B0DRHKXRCQ

AI Demystified: The Fundamentals and Beyond offers a comprehensive exploration of artificial intelligence (AI), presenting its foundational principles, technological advances, and real-world applications while addressing its ethical implications and future potential. This book is designed for readers seeking to understand AI's transformative role in society and its implications across industries. AI Demystified: The Fundamentals and Beyond balances technical depth with accessibility, offering readers a nuanced understanding of AI's current landscape and future trajectory.

Demystifying Quantum Computing: The Future Of Technology Explained

https://www.amazon.com/dp/B0DRN231T8

Quantum computing is poised to reshape the world as we know it. Demystifying Quantum Computing takes readers on an insightful journey into this fascinating field, exploring its foundational principles, transformative applications, and the ethical and societal implications of this revolutionary technology. Through a structured approach and drawing upon authoritative references, the book provides both novices and experts with a clear understanding of quantum computing's potential.

Demystifying Quantum Computing serves as both an educational guide and a forward-looking exploration of quantum computing. By breaking down complex topics and presenting them in a structured, accessible way, it bridges the gap between theoretical understanding and practical application, ensuring readers are prepared for the quantum era.

The Rise Of Intelligent Machines: Ai In Everyday Life

https://www.amazon.com/dp/B0DRHJBB6V

The Rise of Intelligent Machines: AI in Everyday Life is a comprehensive exploration of how artificial intelligence (AI) is transforming various aspects of daily existence. This book delves into the profound impact of AI across multiple domains, offering readers a blend of historical perspective, current applications, and future possibilities.

Throughout the book, optimism is balanced with caution, emphasizing the need for responsible development and governance of AI systems. With engaging narratives and thoughtful analyses, The Rise of Intelligent Machines offers readers a deep understanding of AI's role in shaping the present and the future, equipping them to participate in informed discussions about this transformative technology.

Applications Of Artificial Intelligence

https://www.amazon.com/dp/B0DQW1MJP7

Applications of Artificial Intelligence is a published book series on the applications of artificial intelligence to various industries. Each book dives into a particular industry, the applications of artificial intelligence to that industry, expanding on a few applications, the benefits and challenges for the adoption of artificial intelligence, and the future directions within that industry.

Emergent Minds: Unraveling The Agency In Artificial Intelligence

https://www.amazon.com/dp/B0DRZ4DKJF

In Emergent Minds: Unraveling the Agency in Artificial Intelligence, the convergence of cutting-edge technology, ethics, and philosophy is brought to the forefront. This book delves into the concept of agency within artificial intelligence (AI)—its emergence, implications, and the societal challenges it presents. Drawing from seminal works in AI, ethics, and governance, it offers a comprehensive exploration of how AI systems are evolving beyond tools into entities with autonomous, agent-like behavior.

Target Audience
This book is tailored for professionals in AI, policymakers, ethicists, and curious readers seeking to understand the profound changes AI agency could bring. It bridges technical depth with philosophical inquiry, offering both expert insights and accessible explanations.

Emergent Minds challenges readers to think critically about the evolving role of AI in society. It is both a roadmap for navigating the present and a vision for shaping a future where AI and humanity coexist responsibly.

Quantum Solutions: Practical Applications Across Industries

https://www.amazon.com/dp/B0DRT17T5D

Quantum computing is no longer a futuristic concept—it is a transformative technology already making waves across diverse industries. Quantum Solutions: Practical Applications Across Industries explores the foundational principles of quantum computing and its practical applications, providing a roadmap for professionals, researchers, and decision-makers to harness its power.

The book begins with a clear and accessible introduction to quantum computing, unpacking core concepts like qubits, superposition, entanglement, and quantum gates. It then transitions into industry-specific applications, illustrating how quantum computing is being used to tackle real-world challenges.

Why This Book Matters:
Quantum Solutions: Practical Applications Across Industries serves as a bridge between quantum theory and its practical implications. By blending rigorous academic research with actionable insights, it equips readers to navigate the quantum frontier with confidence. Whether you're an industry leader, a researcher, or simply curious about the quantum revolution, this book offers a comprehensive guide to understanding and leveraging quantum computing in the modern world.

With quantum technology poised to redefine industries and address global challenges, this book provides an essential foundation for embracing the quantum future.

www.ingramcontent.com/pod-product-compliance
Lightning Source LLC
Chambersburg PA
CBHW040221220526
45473CB00001B/73